Eyes on the Street

Police Use of Body-Worn Cameras

in Miami-Dade County

LISA STOLZENBERG

STEWART J. D'ALESSIO

JAMIE L. FLEXON

ACKNOWLEDGMENTS

This project was supported by Grant No. 2015-DJ-BX-K037 awarded by the Bureau of Justice Assistance. The Bureau of Justice Assistance is a component of the Department of Justice's Office of Justice Programs, which also includes the Bureau of Justice Statistics, the National Institute of Justice, the Office of Juvenile Justice and Delinquency Prevention, the Office for Victims of Crime, and the SMART Office. Points of view or opinions in this document are those of the authors and do not necessarily represent the official position or policies of the U.S. Department of Justice.

CONTENTS

1 HIGHLIGHTS

This study evaluates the impact of body-worn camera (BWC) implementation in seven districts served by the Miami-Dade Police Department (MDPD). Nine hundred and sixty-two police officers were provided with BWCs and trained in their use over a four-month period, from May 5, 2016 through August 8, 2016. Statistical analyses of pre- and post-BWC time periods revealed that the use of BWCs resulted in a 34% decrease in the number of citizen complaints against police officers, a 19% reduction in the number of cases of physical response to citizen resistance by police officers, and a 74% drop in the number of civil cases against the MDPD linked to excessive police use of force. These reductions were all statistically significant. Post-BWC results also revealed a decline in the number of internal affair cases, unauthorized force cases, officer injury cases, serious reported crime cases, felony arrests, and civil claims paid by Miami-Dade County for police use of excessive force; however, it cannot be definitively determined, based on available data, that these reductions were due to the use of BWCs rather than to preexisting downward trends. There is scarce empirical evidence to suggest that the use of BWC-based video footage, which can be entered into evidence as additional proof of an alleged incident in court, aids in the prosecution of criminal cases in Miami-Dade County. Finally, the use of BWCs does not appear to have a negative impact on roadway safety by attenuating proactive traffic enforcement: while the number of traffic accidents increased substantially following the implementation of BWCs, there was no significant change in the number of police traffic stops.

2 BACKGROUND

Police frequently interact with the public while performing their daily law enforcement duties. In 2011, over 62.9 million U.S. residents aged 16 or older, or 26% of the population, claimed to have had one or more contacts with police during the prior 12 months. Nine out of 10 citizens reported that they felt the police had acted properly during their encounter. Nonetheless, allegations of police misconduct do arise from civilian-police interactions. Many of these allegations refer to police use of excessive force, with about 8% of excessive force cases culminating in civilian fatality (Packman, 2011).

Growing concern over widespread police misconduct is creating friction between police and the public, especially among minority groups (Weitzer & Tuch, 2004). Surveys consistently indicate that minorities distrust the police more than White citizens do. An analysis of combined data from Gallup polls conducted between 2011 and 2014 reveals that 59% of White citizens have significant confidence in the police, compared to only 37% of Black citizens (Newport, 2014). Surveys also show that Black and Latino citizens are much more likely than White citizens to believe that police unfairly target minorities (Schneider, 2015). A majority of police officers also believe that their interactions with Black citizens have become more aggressive in recent years (Morin, Parker, Stepler, & Mercer, 2017). Minorities' distrust hinders the ability of police to perform their law enforcement duties in an effective manner (U.S. Department of Justice, 2002).

Not only are relations between many citizens and the police deteriorating, but cities are also expending large amounts of money to settle lawsuits against the police for alleged misconduct. The city of Baltimore has paid out an estimated $5.7 million in settlements and another $5.8 million in legal fees since 2011, a figure that likely would have been higher if not for mandatory caps on such payments (Puente, 2014). The city of Chicago paid out approximately $521 million over a 10-year period, through April of 2011. In 2011, Oakland paid out $74 million (Hyatt, 2014), Los Angeles paid out $54 million or about $14 per citizen, and New York City paid out $735 million or $81 per citizen (Goldman, 2012).

It is frequently asserted that BWCs can serve as a viable and cost-effective means to promote both accountability and transparency when police officers interact with citizens. There are various potential benefits to implementing a BWC program in a community: (a) improving the relationship between police and the community at large, as well as minorities in particular; (b) defusing tensions during police-citizen encounters; (c) reducing the number of use-of-force incidents; (d) protecting police officers from unfounded citizen complaints; (e) improving behaviors among members of both law enforcement and the general public; (f) increasing transparency and public perception of police legitimacy; (g) preserving and improving the quality of evidentiary data for use in current and future investigations, arrests, and prosecutions; (h) expediting the resolution of public complaints and lawsuits; (i) enhancing officer safety; (j) improving officers' ability to document and review data used in internal reports and courtroom presentations; and (k) offering law enforcement a tool for self-critique and field evaluation during officer training (Miller, Toliver, & Police Executive Research Forum, 2014).

In 2015, the potential benefits of BWCs led to the establishment of the U.S. Department of Justice's BWC Partnership Program, part of a $263 million three-year initiative instituted by the federal government to enhance community policing (U.S. Department of Justice, 2015). The program's objective was to outfit police officers across the country with 50,000 BWCs through an investment of approximately $75 million. The program requires a 50% investment match by the awardee. In 2015, the MDPD was awarded a $1.2 million grant to purchase and deploy approximately 1,000 BWCs,[1] and it made a matching contribution of $1.2 million to supplement the award.

3 MIAMI-DADE POLICE DEPARTMENT

Miami-Dade County serves an ethnically and racially diverse population of over 2.6 million people in 34 municipalities and an unincorporated area. It is the largest county in the Southeastern United States and the eighth largest in the nation by population. Sixty-six percent of the population is Hispanic, 19% is Black or African American, and 15.2% is non-Hispanic White. Seventy-two percent of residents speak a language other than English at home.[2] Miami-Dade County is anchored by a major seaport and an international airport; it is considered the North American gateway to Latin America and the Caribbean.

The MDPD is currently the eighth largest police organization in the Nation and the largest within the Southeastern United States. With approximately 3,000 sworn personnel, it provides public safety services to more than 2.5 million residents, 13.5 million overnight visitors to Miami-Dade County, and an unincorporated jurisdiction larger than even the largest municipal law enforcement agency within a county, spanning approximately 2,000 square miles. The MDPD is a national leader in law enforcement and is accredited by the Commission of Accreditation for Law Enforcement Agencies (CALEA) and by the Commission for Florida Law Enforcement Accreditation, Inc. It is supported by over 20 specialized bureaus and provides law enforcement services to eight police districts throughout Miami-Dade County: the Airport District, Midwest District, Northside District, South District, Northwest District, Intracoastal District, Kendall District, and West District. These districts are larger in terms of staff and residents than most local municipal departments.

The MDPD has a long history of collaborating with municipal law enforcement agencies, State of Florida law enforcement agencies, the State's Attorney's Office, the Office of the Public Defender, the Miami-Dade County Association of Chiefs of Police, the Florida Department of Law Enforcement, the Miami-Dade Community Relations Board, and the Miami-Dade Juvenile Services Department. The MDPD provides countywide specialized police services in cases of homicide, robbery, narcotics, domestic crimes, and sexual crimes, among others; it also handles crime scene investigations, tactical operations, aviation patrol, special events, crime labs, warrants, court services, central records, and communications. The MDPD uses a variety of technological systems, including automated arrest systems, automated offense incident reports, e-crash technology, automated crime analysis tools, crime mapping, electronic document management, automated fingerprint identification, eNotify software for automated subpoena notifications, ePars software for automated payroll processing, and an event scheduling system.

As a world leader in a variety of investigative and enforcement disciplines, the MDPD is actively expanding its Real-Time Crime Center to meet twenty-first century challenges. The center is fully funded and supported by both agency leadership and county government. Its highly experienced intelligence analysts undergo regular training and are given access to every existing database so as to meet their investigative goals. The MDPD also makes use of a crime data warehouse, a crime analysis system, a sexual crime analysis system, eCitation software for electronic ticketing, ePolice software for call management, Uniform Crime Reports (UCRs), field interviews, computer-aided dispatches, the DHS-recognized Southeast Florida Fusion Center, rapid identification for wireless mobile fingerprint capture, a mugshot system, and an online citizen crime reporting system.

4 BODY-WORN CAMERAS

In many large metropolitan communities such as Miami-Dade County, racial tensions have often escalated between law enforcement and minority groups, leading to widespread mistrust of the police. On November 25, 2014, in response to a Missouri grand jury's decision not to charge the police officer who fatally shot 18-year-old Michael Brown, approximately one hundred protestors rallied outside Miami-Dade's criminal courthouse (Richard E. Gerstein Justice Building). On December 5, 2014, in response to the deaths of several unarmed men at the hands of law enforcement officers, hundreds of protestors shut down a five-mile stretch of highway I-95 and the causeway near Downtown Miami, chanting "Hands Up, Don't Shoot." Miami-Dade Police Department resources were deployed to assist in policing both events.

The MDPD BWC project was a concerted effort to address general mistrust of law enforcement in Miami-Dade County, resulting from national, regional, and local misuse of police force. The project's overriding goal is to improve public relations by increasing transparency and accountability in police-citizen encounters. The project was managed by two teams. One team met regularly with criminal justice system stakeholders to direct the project, address all policy and performance issues, deliverables' status, outcome measurements, and research findings. The second team focused on the implementation of the BWCs and recommended corrective actions for non-compliance and problematic field-identified issues. The BWC project's specific objectives and implementation plan are described below.

The project's first objective was to establish privacy policies and operational procedures that were transparent, accessible to the public, and mindful of BWC issues involving Freedom of Information Act (FOIA) liabilities, civil rights, domestic violence, juvenile groups, and victim groups. Prior research conducted by the MDPD, as well as communication with law enforcement agencies around the country already using BWCs, indicated that several policies should be in place for the successful deployment of BWCs in Miami-Dade County. These policies include clear and comprehensive BWC guidelines, agreements with the State Attorney, the Public Defender, and the Chief Judge on how BWC video footage should be used in court proceedings and disseminated as part of discovery, and a clear retention schedule for evidentiary and non-evidentiary BWC data in order to keep storage costs down.

The MDPD developed and implemented appropriate privacy policies that addressed issues such as the use of BWCs inside civilian homes, the establishment of an activation policy for appropriate use of BWCs, the guarantee of public release and redaction policies complying with Florida public records laws, and the alignment of retention schedules with those published by Florida's Division of Library and Information Services. Program policies and procedures address the use, review, access, storage, labeling, retention, redaction, and expungement of all digital video evidence. All BWC system data, including data on police-related shootings, are the sole property of the MDPD and can be used for official purposes only. The MDPD's video forensics unit is responsible for maintaining all BWC system data. Body-worn camera activation commences immediately before police officers exit their vehicles or as soon as practicable when responding to a call for service or an official law enforcement matter. In locations such as residences, where victims have a reasonable expectation of privacy, an officer may honor a victim's request to turn off the BWC unless the recording is being made pursuant to an arrest or search of the residence or individuals situated therein. When a BWC is used in an investigation, video footage is documented in an Offense/Incident Report, a Field Interview Report, or a uniform traffic citation. All other citizen contacts captured by a BWC are recorded in the corresponding officer's Daily Activity Report. The

MDPD obtained input from a variety of community groups to ensure that privacy issues associated with juveniles, victims, and domestic violence cases were considered when crafting BWC policy.

The project's second objective was to engage broad stakeholders, local politicians, and community members in order to address policy, training, deployment, and procurement-related requirements. Through the participation of the county mayor and the board of county commissioners, ongoing discussions with numerous community and stakeholder organizations, and three years of extensive research and market analyses, the MDPD was able to draft a program policy and a procurement solicitation that was shared with the community and stakeholder agencies for feedback. This allowed the MDPD to implement the BWC project throughout Miami-Dade County in an expeditious manner. Miami-Dade County competitively solicited proposals from qualified firms to provide a turnkey, cloud-based BWC and video management system that could capture footage from a law enforcement officer's perspective and store all recorded material in a secure hosted website. Services provided by the selected contractor during the contract term included installation of all BWCs and associated video management software; host video/data storage; configuration, implementation, and ongoing maintenance support services; and training services for MDPD staff. The BWC technology complied with all recommended minimum operating features identified in the solicitation, except for those related to night-time/low-light functionality. Notably, video footage captured by the BWCs closely resembles what a police officer observes, without alterations or enhancements that can be misconstrued.

The project's third objective was to train law enforcement personnel on BWC privacy policies and operational procedures. The selected contractor was responsible for training 15 MDPD system administrators, 60 train-the-trainers, 100 command staff, and 1,500 officers on the proposed BWC and video management system. All BWC operators were required to receive hands-on training prior to gaining full access to the system and must complete an annual refresher training course provided by the Miami-Dade Public Safety Training Institute. Initial training consisted of an eight-hour session split into four hours of vendor-facilitated instruction on the use of BWCs and four hours of MDPD-facilitated training on the corresponding policies and procedures. The annual refresher training course consists of an updated review of policies, procedures, best practices, and BWC use.

The project's fourth objective was to increase BWC knowledge and build criminal justice practitioners' problem-solving capacity by allowing access to digital multimedia evidence.

Body-worn camera recordings are digitized, uploaded, and subsequently shared with different criminal justice system agencies via a cloud-based system. To assist criminal justice practitioners, the MDPD facilitated training sessions, forums, and focus groups with partner agencies on the BWC project's privacy policies, operational procedures, and cloud-based system procedures for access to and use of BWC recordings as digital multimedia evidence. Miami-Dade Police Department trainers worked with the State Attorney's Office, the Office of the Public Defender, and the Administrative Office of the Courts to coordinate these sessions. Additionally, the MDPD created a user's guide for criminal justice practitioners and attorneys, made available upon request, on the capabilities and limitations of the device.

5 SUMMARY

Despite the potential benefits of BWCs, there are few studies investigating their effect on police behavior. Moreover, much of the research on BWCs conducted to date is methodologically problematic, making its conclusions tentative at best. The first problem is that most previous evaluations of BWCs' effectiveness in reducing violence against police rely on small samples, because few police officers were equipped with BWCs when these studies were conducted. To illustrate, Headley, Guerette, and Shariati (2017) looked at 26 officers equipped with BWCs, Farrar (2013) looked at 54, Katz, Choate, Ready, and Nuño (2014) looked at 56, the Mesa Police Department (2013) looked at 50, Ariel, Farrar, and Sutherland (2015) looked at 54, and ODS Consulting (2011) looked at 39. On the basis of these small samples, it has been found that there are approximately 10.9 use-of-force allegations per 100 full-time sworn officers at large police agencies responding to calls for service (Hickman, 2006). The small number of police officers equipped with BWCs, coupled with the small number of use-of-force incidents per police officer, makes it exceedingly difficult to derive stable estimates on the impact of BWCs.

Moreover, previous studies are procedurally confounded because of how police officers actually operate in the field. In these studies, police officers were randomly assigned to experimental and control groups. Officers in the experimental group were given BWCs, while officers in the control group were not. The two groups were then compared regarding citizen complaints, use-of-force incidents, etc. The problem with this methodological approach is that police officers with BWCs often respond to the same calls for service as police officers without BWCs. Research shows that, at least in heavily populated cities, it is relatively common for multiple police officers to interact with citizens. For example, a survey of youth stopped by police in Chicago indicated that 67.4% of respondents were confronted by multiple police officers rather than by a police officer acting alone.[3]

Finally, prior research often resorted to a quasi-experimental design to evaluate the effectiveness of BWCs. For example, Ariel et al. (2015) created an interrupted time series autoregressive integrated moving average (ARIMA) model for only 12 pre-intervention monthly time periods. While an interrupted time series ARIMA model is appropriate for interpreting aggregate change, at least 48 pre-intervention periods are required for trend and seasonality to be accurately modeled (McCleary & Hay, 1980); otherwise, ARIMA results are unstable (McCain & McCleary, 1979).

The current study aims to advance extant literature on BWC use by analyzing longitudinal data to ascertain whether the implementation of BWCs in Miami-Dade County has impacted law enforcement, crime, prosecution, and civil litigation. The influence of BWCs on proactive traffic enforcement is also assessed. The policymaking implications of this study are significant. If BWCs are effective in curtailing citizen complaints, internal affair investigations, use-of-force incidents, etc., then their use may enhance public safety. However, if BWC use has a negligible impact in these areas, then alternative strategies to improve public safety would need to be identified and designed.

6 DATA AND METHODS

The data analyzed were obtained from the MDPD and include the seven districts served by the department from January 1, 2005 to June 30, 2018 (the Airport District was excluded). As indicated in Table 1, 962 police officers in all seven districts received BWCs and were trained in their use over a four-month period. As of June 30, 2018, the MDPD had allocated BWCs to 1,869 officers throughout Miami-Dade County, including 991 officers in the seven police districts.

Tables 2 and 3 describe the outcomes analyzed in this study. Most outcome data were available for 162 monthly time periods. The pre-BWC time interval ranges from January 2005 to April 2016 (i.e., 136 months), while the post-BWC time interval ranges from May 2016 to June 2018 (i.e., 26 months). Two outcomes were reported annually: the number of claims made against the MDPD and the number of claims paid by Miami-Dade County for police actions involving use of force. The pre-BWC time interval for these annual outcomes ranges from 2005 to 2015 (i.e., 11 years), while the post-BWC time interval ranges from 2016 to

2017 (i.e., 2 years).

Outcomes were first evaluated through descriptive statistics—in particular, through the calculation of their mean value before and after the implementation of BWCs in May 2016. For comparative purposes, mean values reflect an equal number of pre- and post-BWC time periods (i.e., 26 months). However, it is important to stress that simple comparisons of pre- and post-BWC mean values are suggestive at best, since changes in any given outcome could be due to a preexisting trend. An interrupted time series ARIMA model was therefore used to evaluate the impact of BWC implementation on all monthly outcomes. This statistical procedure is effective in drawing causal inferences, as it helps to eliminate a large number of plausible alternative explanations for a hypothesized causal relationship through statistical comparison of pre- and post-intervention series (Johnson & Christensen, 2008).

As previously mentioned, at least 48 pre-intervention periods are needed to accurately model trend and seasonality through interrupted time series ARIMA analysis (McCleary & Hay, 1980). Thus, while ARIMA was used to analyze monthly outcomes, yearly outcomes were analyzed using a One-Way ANOVA procedure.

TABLE 1. IMPLEMENTATION OF THE BODY-WORN CAMERAS IN MDPD DISTRICTS

District	Dates BWC distributed during implementation		Officers equipped	
			During implementation	As of 6/30/18
Midwest	5/2/16	5/19/16	124	135
Northside	5/24/16	6/3/16	167	173
South	6/7/16	6/16/16	163	170
Northwest	7/25/16	8/10/16	87	87
Intracoastal	7/25/16	8/10/16	129	126
Kendall	8/15/16	8/29/16	132	134
Hammocks	8/15/16	8/29/16	160	166
TOTAL			962	991

TABLE 2. DEFINITIONS OF OUTCOMES ANALYZED

Outcome	Description
Contact reports	Number of citizen complaints filed against police officers per month.
SRRR	Number of Supervisor Response to Resistance Reports received per month. A police officer files this report when he or she determines it is necessary to respond to resistance to control a situation and to prevent possible harm to the officer or others. When these situations occur, department policy requires the primary reporting officer to write the initial incident report and his or her supervisor to review the report.
Internal affairs	Number of internal affairs cases relating to the investigation of public complaints against police officers per month.
Unauthorized force	Number of police use of unauthorized force cases per month.
Officers injured	Number of police officers injured during a documented use of force situation per month.
Part 1 crimes	Number of Part 1 crimes reported in Miami-Dade County per month. Part 1 crimes include murder and nonnegligent manslaughter, forcible sex offenses, robbery, aggravated assault, burglary, larceny, and motor vehicle theft.
Felony arrests	Number of individuals arrested for one or more felony offenses in Miami-Dade County per month.

Outcome	Description
Felony charges filed	Number of felony charges filed in Miami-Dade County per month.
Felony guilty pleas	Number of felony guilty pleas in Miami-Dade County per month.
Felony guilty verdicts	Number of felony guilty verdicts in Miami-Dade County per month.
Civil cases filed	Number of civil cases filed against MDPD by calendar year for the actions of officers involved in shootings, use of force, and other actions related to use of force.
Claims paid	Claims paid by Miami-Dade County by calendar year for the actions of MDPD officers involved in shootings, use of force, and other actions related to use of force.
Traffic stops	Number of traffic stops per month.
Traffic accidents	Number of traffic accidents per month.
Hit and runs	Number of hit and runs per month.
Traffic fatalities	Number of traffic fatalities per month.

TABLE 3. DESCRIPTION OF OUTCOMES ANALYZED (JANUARY 2005-JUNE 2018)

	Pre-BWC		Post-BWC	
Outcome	# time periods	Mean (SD)	# time periods	Mean (SD)
Contact reports	136	18.27 (5.35)	26	12.73 (6.06)
SRRR	136	28.19 (5.18)	26	20.69 (5.44)
Internal affairs	136	16.47 (6.65)	26	7.04 (2.51)
Unauthorized force	136	6.21 (3.08)	26	2.42 (1.33)
Officers injured	136	6.99 (3.56)	26	5.42 (2.42)
Part 1 crimes	136	4,059.79 (477.39)	26	3263.77 (253.16)
Felony arrests	136	3,083.16 (639.06)	26	2,022.81 (179.95)
Felony charges filed	136	3,507.76 (895.16)	26	2,147.61 (348.59)
Felony guilty pleas	136	1,856.52 (551.35)	26	709.65 (236.50)

24

	Pre-BWC		Post-BWC	
Outcome	# time periods	Mean (SD)	# time periods	Mean (SD)
Felony guilty verdicts	136	2,206.05 (558.78)	26	1,334.61 (201.00)
Civil cases filed	11	31.45 (9.49)	2	8.00 (5.66)
Claims paid	11	$959,629.54 (482,289.25)	2	$675,156.37 (360,408.77)
Traffic stops	136	5,988.59 (1,727.60)	26	3,153.54 (488.43)
Traffic accidents	136	3,218.78 (270.97)	26	3,686.31 (266.70)
Hit and runs	136	664.06 (71.60)	26	742.38 (59.42)
Traffic fatalities	136	7.25 (2.88)	26	7.27 (2.97)

7 LAW ENFORCEMENT OUTCOMES

It is widely believed that the use of BWCs helps to reduce the number of citizen complaints filed against police officers (contact reports), Supervisor Response to Resistance Reports (SRRR), internal affair investigations of police officers, unauthorized use-of-force cases, and cases of assaultive violence against police officers. To illustrate, the Urban Institute's (Peterson, Yu, La Vigne, & Lawrence, 2018) evaluation of the Milwaukee Police Department's BWC program found that complaints were made against 9.52% of officers with BWCs and 12.7% of officers without BWCs. The Police and Crime Standards Directorate (2007) found that the number of citizen complaints against police officers declined by 14.3% overall after the implementation of BWCs. ODS Consulting (2011) found that only seven complaints were made against officers wearing BWCs in both Renfrewshire and Aberdeen, out of more than 5,000 contacts registered during the study period. These findings are compatible with further research. Ariel et al. (2015) found an 88% decline in the number of citizen complaints following the implementation of BWCs in Rialto,

California. In another study undertaken in Spokane, Washington, researchers found that there were 28% fewer complaints against police officers with BWCs than against officers without BWCs (White, Gaub, & Todak, 2018).

Studies have also revealed that BWCs deter unwarranted use-of-force incidents while contributing to citizen compliance with police directives. For example, Ariel et al. (2015) observed a 60% reduction in the number of use-of-force incidents following the implementation of BWCs. Similarly, Henstock and Ariel (2017) examined the use of BWCs among British police split into camera and no-camera groups during 430 shifts. They found a 35% decrease in the number of use-of-force incidents among BWC groups. This was mostly due to a decrease in the number of police actions involving physical restraint of the defendant or non-compliant handcuffing, rather than aggressive actions such as use of Tasers or pepper spray. Work by Jennings, Lynch, and Fridell (2015) is also consistent with the above findings: researchers found that the use of BWCs results in a significant decrease in the number of serious citizen complaints and response-to-resistance incidents. Notably, research also found a reduction in the number of substantiated citizen complaints against police officers when BWCs were used (Katz et al., 2014).

Findings suggest that coercive police behavior against citizens needs to be attenuated because it produces antagonism and a loss of police legitimacy in the eyes of the public, particularly among minority citizens (Holmes, 2000). Said antagonism, coupled with the frequent lack of respect towards police officers displayed by some members of the community, may encourage other citizens to become belligerent when interacting with police; it may further motivate them to assault police officers. This was clearly demonstrated on July 16, 2016 in Dallas, Texas, when five police officers were killed and nine other officers were injured during an ambush. The attack took place during a peaceful protest against the killing by police of Black males Alton Sterling of Baton Rouge, Louisiana and Philando Castile of Falcon Heights, Minnesota. This and other attacks (e.g., Pittsburg, PA) demonstrate that police behavior, even when taken out of context, may prompt certain individuals to partake in criminal activities (Kirk & Matsuda, 2011; Kirk & Papachristos, 2011; Sherman, 1993).

Some speculate that the reduction in cases of police use of force and assaultive violence directed at police post-BWC implementation responds to a "civilization effect" (White, 2014). People are more apt to behave in a "socially acceptable" manner when an outside third party, who can impose either informal or formal sanctions, is monitoring their interactions with others. This modification in behavior is referred to as "reactivity" and was originally documented in the Western Electric Company's worker productivity studies (Landsberger, 1958). When researchers increased lighting at the Hawthorne Works factory in Cicero, Illinois, worker productivity increased as expected. However, when lighting was decreased, worker productivity unexpectedly increased as well. Further research revealed that the latter resulted from workers recognizing that their activities were being scrutinized: because they believed that they were being monitored by a third party who could adversely influence their employment at the factory, and because being productive was a social expectation, workers were motivated to behave in a manner that increased their productivity notwithstanding the status of lighting. This effect has also been noted in survey-based research: people often feel obligated to respond to questions in a manner consistent with prescribed norms or values (Cox, Jones, & Navarro-Rivera, 2014). Researchers have further found that individuals are much less apt to be abrasive in their interaction with others over the Internet when their identity is known, given their fear of being socially ridiculed or ostracized (Suler, 2004). Based on prior research, then, one might argue that BWCs reduce anonymity when police officers and citizens interact, and that therefore both parties will act more civilly towards each other because civil behavior is socially expected.

As illustrated in Figure 1, the number of contact reports declined by 34% following the implementation of BWCs in Miami-

Dade County's seven districts (pre-intervention monthly mean = 19.3 vs. post-intervention monthly mean = 12.7). Furthermore, the number of SRRRs decreased by 19% (pre-intervention monthly mean = 25.7 vs. post-intervention monthly mean = 20.7).

FIGURE 1. AVERAGE MONTHLY LAW ENFORCEMENT OUTCOMES

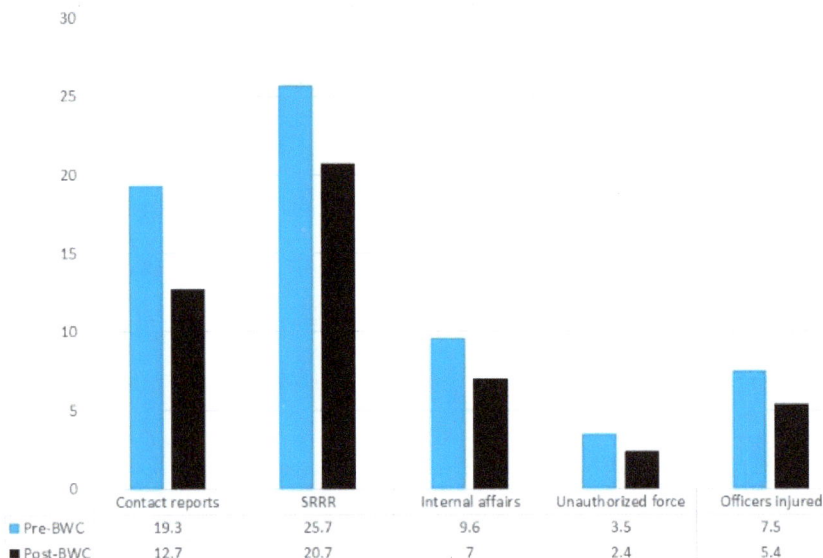

	Contact reports	SRRR	Internal affairs	Unauthorized force	Officers injured
Pre-BWC	19.3	25.7	9.6	3.5	7.5
Post-BWC	12.7	20.7	7	2.4	5.4

Figure 1 also depicts mean changes, between pre- and post-intervention periods, in the number of internal affair investigations, unauthorized use-of-force cases, and police officer injuries during documented use-of-force situations. The mean monthly number of internal affair investigations was 9.6 pre-intervention and 7.0 post-intervention. The overall mean monthly number of unauthorized use-of-force cases and injured officers was, respectively, 3.5 and 7.5 pre-intervention and 2.4 and 5.4 post-intervention. Only mean monthly values for the number of contact reports and SRRRs decreased significantly between pre- and post-intervention periods, according to interrupted time series analyses.

8 REPORTED CRIMES AND FELONY ARRESTS

ODS Consulting (2011) found that, in certain areas within two Scottish communities (Renfrewshire and Aberdeen) in which BWCs were deployed, the number of police offenses declined by 19%, acts of vandalism by 29%, minor assaults by 27%, and serious assaults by 60%, for an overall decline in crime of 26%. Other studies also report that the implementation of BWCs attenuates criminal activity. It is plausible that this decrease in crime rates, as well as an increase in the number of arrests, is due to greater citizen confidence in the police post-BWC implementation.

Greater public confidence in the police may reduce crime rates by decreasing the likelihood that citizens will take the law into their own hands to resolve personal disputes; the latter, known as "self-help," accounts for a considerable portion of socially violent situations (Black, 1983). Cases of self-help occur when people believe that the police are ineffective or unwilling to help them resolve their grievances with others. Violence escalates when other citizens in the community react aggressively to the heightened possibility of a violent encounter (Peterson, Krivo, & Browning, 2006). Evidence supporting this position can be gleaned from surveys indicating that firearm ownership figures are substantially higher among citizens who lack confidence in the police (Kleck, 1997). Thus, it seems plausible that the implementation of BWCs reduces criminal activity rates by increasing citizen confidence in the police.

Figure 2 shows the number of reported Part 1 crimes and felony arrests over time. Examination reveals that the mean monthly number of reported crimes and arrests decreased by 6.6% and 10.8%, respectively, following BWC implementation. However, their decrease was not salient per ARIMA analyses.

FIGURE 2. MONTHLY PART 1 CRIMES AND FELONY ARRESTS

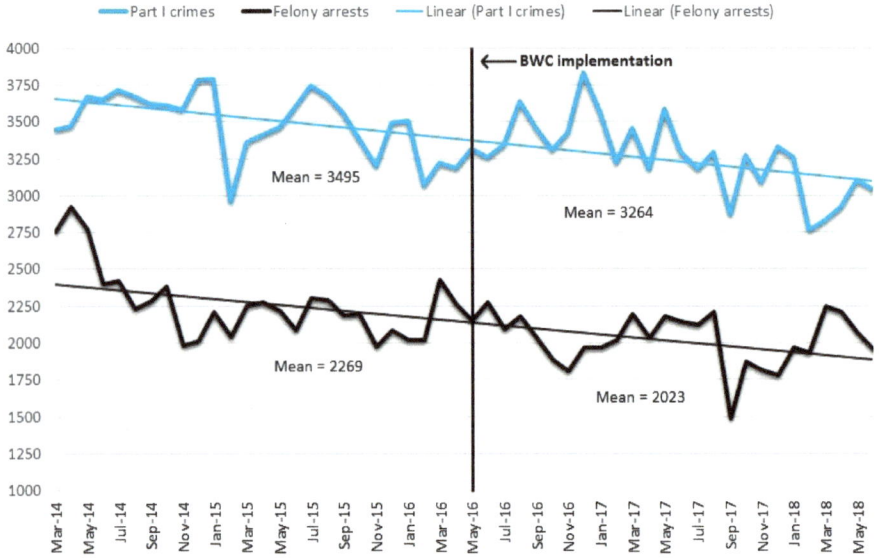

9 FELONY CHARGES, GUILTY PLEAS AND VERDICTS

The use of BWCs may increase the number of filed felony charges, guilty pleas, and guilty verdicts, because video footage can be entered into evidence as additional proof of an alleged crime. Figure 3, however, reveals a decrease in all three categories following BWC implementation. Specifically, the number of charges filed dropped by 15%, the number of guilty pleas by 38%, and the number of guilty verdicts by 14%. However, according to ARIMA analyses, these changes were relatively insignificant. Thus, it does not appear that BWCs have a positive influence on the prosecution of criminal cases. It is important to point out, however, that cities not subjected to BWC use were included in the analyses because prosecution data for Miami-Dade County are aggregated at the county-level.

FIGURE 3. AVERAGE MONTHLY FELONY CHARGES FILED, GUILTY PLEAS, AND GUILTY VERDICTS

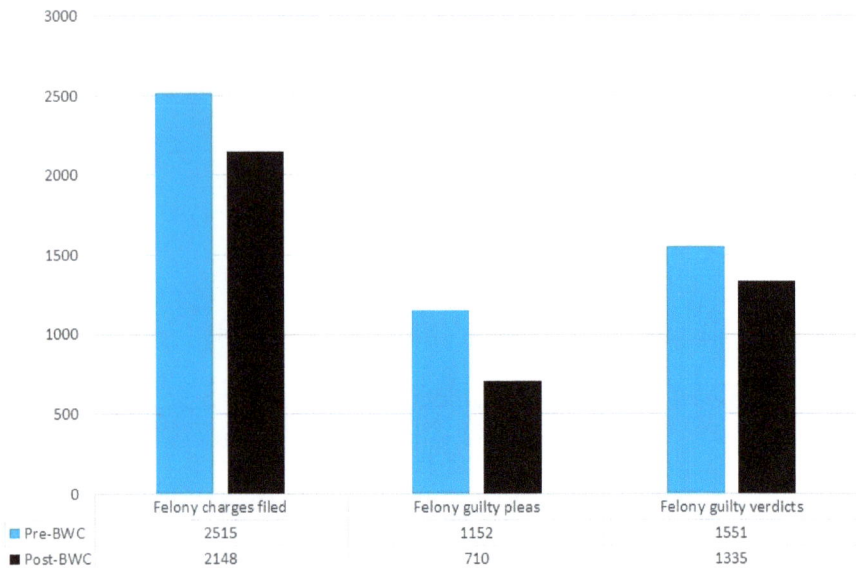

	Felony charges filed	Felony guilty pleas	Felony guilty verdicts
Pre-BWC	2515	1152	1551
Post-BWC	2148	710	1335

10 CIVIL CASES FILED AND CLAIMS PAID

Because BWC video footage can be entered into evidence as additional proof of an alleged criminal incident, BWC use by police has the potential of influencing the prosecution of criminal defendants. Prior research has found that BWCs increase the likelihood of cases ending in a guilty plea rather than a criminal trial, and of trials resulting in a guilty verdict because of video evidence. For example, the Police and Crime Standards Directorate (2007) found that the use of BWCs resulted in an increase in the number of charges and summons, the number of sanction detections, and the number of cases in which a violent incident was deemed a chargeable crime. Katz et al. (2014) observed that prosecuted domestic violence cases were more apt to result in a guilty plea or in a guilty verdict at trial after BWC implementation. Other studies have also found that video footage improves the quality of evidence used in the prosecution of criminal defendants (ODS Consulting, 2011; Morrow, Katz, & Choate, 2016).

Civil litigation against police officers' use of excessive force represents a financial burden for Miami-Dade County. The annual cost incurred by the county to support internal affair investigations based on use-of-force complaints is approximately $2.1 million, including investigating officers' salary and fringe benefits. The ability of BWCs to furnish indisputable video evidence of the circumstances surrounding a use-of-force incident may help to attenuate these monetary costs.

An examination of civil litigation cases involving correctional officers may be instructive. In many of these cases, video footage from closed-circuit television (CCTV) helps to clarify whether inmate injury or death occurs as a result of either justified or excessive use of force by correctional officers. In several cases, officers have been found criminally liable and faced extensive prison time. It is therefore reasonable to assume that video footage of use-of-force incidents will be valuable in either challenging or proving liability. However, because taxpayers rather than offending officers are typically held accountable for civil damages, it remains unclear as to the extent that BWCs will reduce litigation costs.

Figures 4 and 5 show the number of civil cases filed against MDPD and the number of claims paid by Miami-Dade County for MDPD officer actions involving excessive use of force from January 1, 2005 through December 31, 2017. The red shading for 2016 and 2017 indicates that BWCs were deployed. The mean annual number of civil cases filed against MDPD dropped from 31 pre-intervention to 8 post-intervention. This represents a 74% decrease.

The average amount paid by Miami-Dade County for claims also decreased from $959,629 to $675,156 (i.e., 30%). Although the number of civil cases filed against MDPD and the number of claims paid by the county both decreased following the implementation of BWCs, ANOVA analyses showed a significant decline in the number of civil cases filed but no substantive change in the number of claims paid.

FIGURE 4. CIVIL CASES FILED AGAINST MDPD

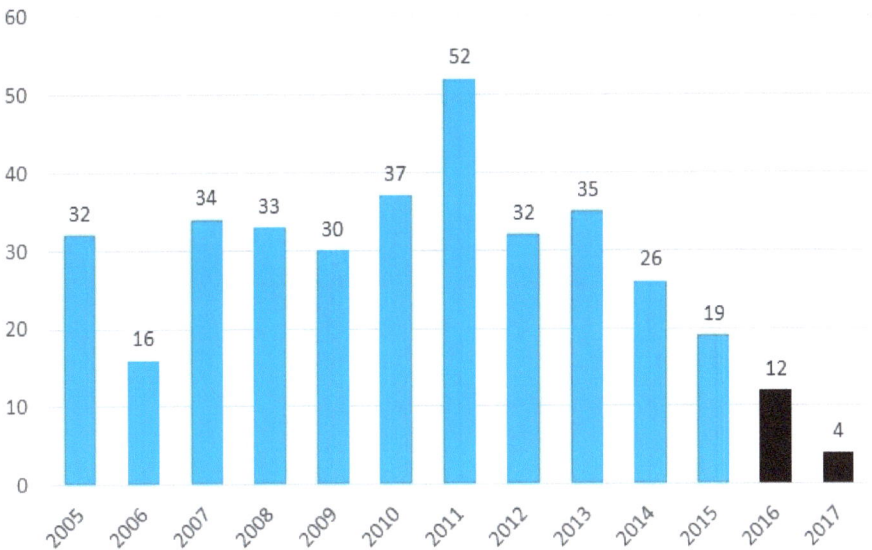

FIGURE 5. CLAIMS PAID BY MIAMI-DADE COUNTY

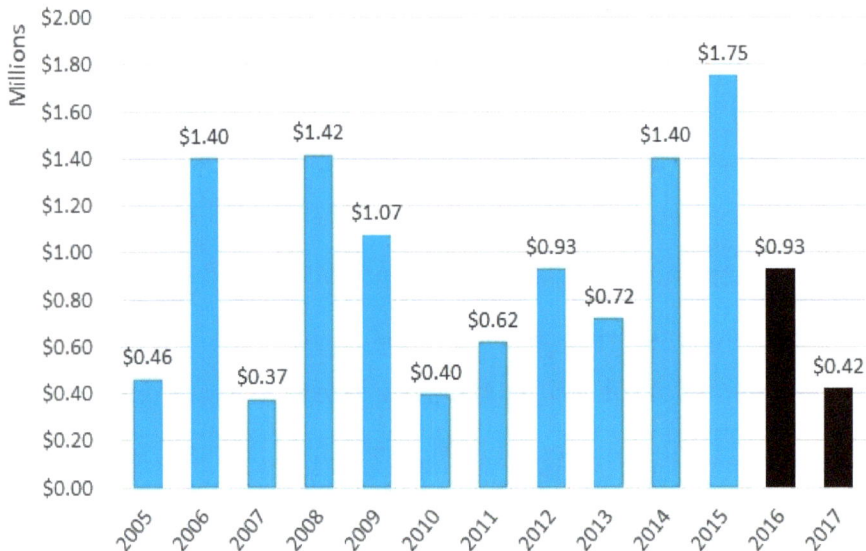

11 TRAFFIC OUTCOMES

While the benefits of BWCs are often touted, there is anecdotal evidence that their use may result in less proactive policing, particularly for police-initiated forms of enforcement such as traffic stops and stop-and-frisk searches. For example, the Urban Institute (Peterson et al., 2018) found that officers with BWCs conducted significantly fewer traffic stops than officers without BWCs. The Mesa Police Department (2013) also found that officers equipped with BWCs conducted significantly fewer stop-and-frisks and arrests than officers who were not wearing cameras. This is problematic, because public safety may be compromised when police make fewer traffic stops (Luca, 2015) or fail to conduct stop-and-frisk searches (MacDonald, Fagan, & Geller, 2016).

Data also show that the number of police-citizen contacts has declined in recent years. In 2008, approximately 5.3 million fewer residents had face-to-face contacts with police than in 2002: figures dropped from 45.3 to 40.0 million (Eith & Durose, 2011). While this decline may be partly due to a reduction in both violent crime rates and property crime rates (Truman & Langton, 2015), it may also be due to police initiating fewer contacts with the public for fear of negative repercussions, a process known as de-policing. A recent national survey conducted by the Pew Research Center found that 72% of queried police officers reported being significantly less willing to stop and question suspicious persons than they used to be, because they currently fear possible negative media attention (Morin et al., 2017). In a speech at the University of Chicago Law School, former FBI Director James Comey lamented the fact that police are becoming less aggressive in the wake of several well-publicized police shootings (Schmidt & Apuzzo, 2015). Some claim that this lackadaisical attitude towards law enforcement, sprouting from fear of public scrutiny, is allowing criminal activity rates to escalate. The phenomenon is often referred to as the "Ferguson effect" (MacDonald, 2015), after a White police officer's fatal shooting of 18-year-old Black male Michael Brown on August 9, 2014 led to considerable civil unrest. Despite the phenomenon's plausibility, the limited amount of research on the topic has not yielded substantial evidence for it (Pyrooz, Decker, Wolfe, & Shjarback, 2016).

Figure 6 depicts the mean monthly number of traffic stops, traffic accidents, hit-and-runs, and traffic fatalities for pre- and post-intervention periods. As can be seen, the number of traffic stops dropped by 14%, while the number of traffic accidents and hit-and-runs rose by 7% and 11%, respectively. There was no change in the number of traffic fatalities. ARIMA results revealed that the rise in the number of traffic accidents did not respond to a preexisting trend. However, because there was no substantive change in the number of traffic stops made by police following the deployment of BWCs, the rise in the number of traffic accidents was most likely not the consequence of less proactive traffic enforcement.

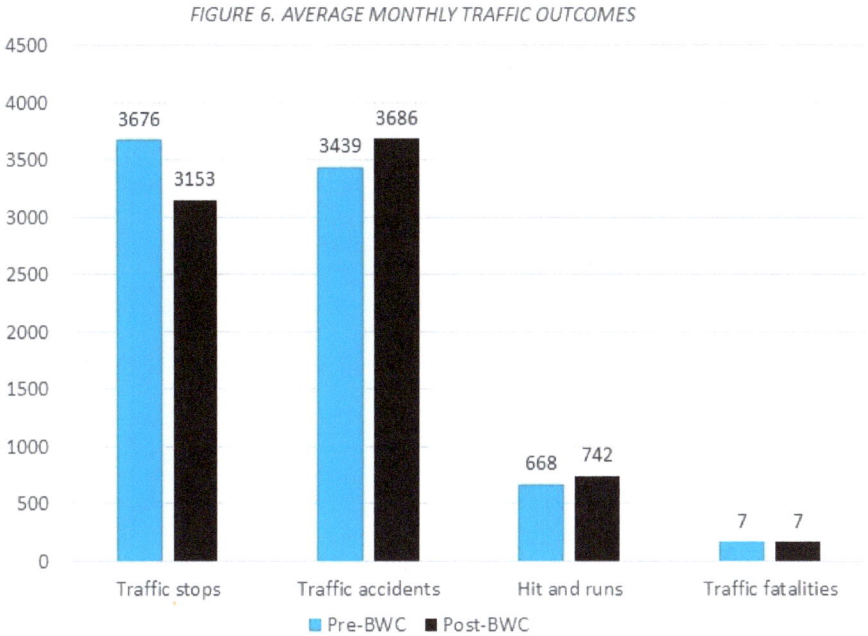

FIGURE 6. AVERAGE MONTHLY TRAFFIC OUTCOMES

12 CONCLUSION

The current study overcomes several pitfalls of previous BWC-related evaluations due to its longitudinal design and large, inclusive sample of subjects: 962 MDPD officers across seven districts were provided with BWCs and monitored from May 5, 2016 through August 8, 2016 in order to study the cameras' impact. Pre- and post-statistical analyses revealed several important findings. The use of BWCs resulted in a 34% decrease in the number of citizen complaints against police officers, a 19% reduction in the number of cases of physical response to citizen resistance by police officers, and a 74% drop in the number of civil cases against the MDPD related to excessive police use of force. These findings are robust and do not appear to be method-dependent, which indicates that the use of BWCs addresses several areas of concern. It also appears to yield a decrease in the number of internal affair cases, unauthorized force cases, officer injury cases, serious reported crime cases, felony arrests, and claims paid by Miami-Dade County for police use of excessive force. However, these findings are less robust owing to truncated data, and future research is encouraged.

The data further suggest that the availability, influence, and impact of BWC video evidence does not aid in the prosecution of criminal cases in Miami-Dade County. As previously discussed, this finding may be an artifact of the data and requires additional vetting, as some cities not subjected to the use of BWCs were included in the analysis due to data aggregation at the county-level.

The use of BWCs does not appear to impact roadway safety. Despite a substantial increase in the number of traffic accidents following the implementation of BWCs, there was no significant change in the number of traffic stops. This finding contradicts the idea that BWCs are having a detrimental impact on public safety in Miami-Dade County as a result of de-policing.

In sum, this study's findings indicate that BWC use by police has the power to significantly influence current and future policy related to police-community relations, police-citizen interactions, and litigation related to police use of force without resulting in de-policing or being a threat to public safety. Research, however, is incipient; replication studies, particularly those with a longitudinal design and large sample sizes, are needed to enhance our understanding of this promising technology.

NOTES

1. This award was made under Funding Category 1: Implementation of New Body-Worn Camera Program for Large Agencies (BJA-2015-4169).

2. Census.gov People *QuickFacts* for Miami-Dade County, Florida.

3. The authors' analysis of data collected by Friedman, Warren, Arthur J. Lurigio, Richard Greenleaf, and Stephanie Albertson. 2004. Encounters between police officers and youths: The social costs. *Journal of Crime and Justice* 27: 1-25.

REFERENCES

Ariel, B., Farrar, W. A., & Sutherland, A. (2015). The effect of police body-worn cameras on use of force and citizens' complaints against the police: A randomized controlled trial. *Journal of Quantitative Criminology, 31*, 509-35.

Black, D. (1983). Crime as social control. *American Sociological Review, 48*, 34-45.

Cox, D., Jones, R.P., & Navarro-Rivera, J. (2014). *I Know What You Did Last Sunday: Measuring Social Desirability Bias in Self-Reported Religious Behavior, Belief, and Identity*. Washington, DC: Public Religion Research Institute.

Eith, C., & Durose, M. R. (2011). *Contacts between Police and the Public, 2008*. Washington, DC: Bureau of Justice Statistics.

Farrar, W. (2013). *Self-Awareness to Being Watched and Socially-Desirable Behavior: A Field Experiment on the Effect of Body-Worn Cameras and Police Use-of-Force*. Washington, DC: Police Foundation.

Friedman, W., Lurigio, A.J., Greenleaf, R., & Albertson, S. (2004). Encounters between police officers and youths: The social costs. *Journal of Crime and Justice, 27*, 1-25.

Goldman, H. (2012). NYPD abuse increases settlements costing city $735 million. *Bloomberg Businessweek*, September 4, 2012.

Headley, A. M., Guerette, R. T., & Shariati, A. (2017). A field experiment of the impact of body-worn cameras (BWCs) on police officer behavior and perceptions. *Journal of Criminal Justice, 53*, 102–09.

Henstock, D., & Ariel, B. (2017). Testing the effects of police body-worn cameras on use of force during arrests: A randomised controlled trial in a large British police force. *European Journal of Criminology, 4*, 720-50.

Hickman, M. J. (2006). *Citizen Complaints about Police Use of Force*. Washington, DC: Bureau of Justice Statistics.

Holmes, M. D. (2000). Minority threat and police brutality: Determinants of civil rights criminal complaints in U.S. municipalities. *Criminology, 38*, 343-68.

Hyatt, A. (2014). *Oakland Spent $74 Million Settling 417 Police Brutality Lawsuits—Oakland Police Beat*. Retrieved from: https://blog.sfgate.com/inoakland/2014/04/14/oakland-spent-74-million-settling-417-police-brutality-lawsuits-oakland-police-beat/

Jennings, W. G., Lynch, M. D., & Fridell, L. A. (2015). Evaluating the impact of police officer body-worn cameras (BWCs) on response-to-resistance and serious external complaints: Evidence from the Orlando police department (OPD) experience utilizing a randomized controlled experiment. *Journal of Criminal Justice, 43,* 480-86.

Johnson, B., & Christensen, L. (2008). *Educational Research: Quantitative, Qualitative, and Mixed Approaches.* Thousand Oaks, CA: Sage.

Katz, C. M., Choate, D.E., Ready, J. R., & Nuño, L. (2014). *Evaluating the Impact of Officer Worn Body Cameras in the Phoenix Police Department.* Phoenix, AZ: Center for Violence Prevention and Community Safety, Arizona State University.

Kirk, D. S., & Papachristos, A. V. (2011). Cultural mechanisms and the persistence of neighborhood violence. *American Journal of Sociology, 116,* 1190-1233.

Kirk, D. S., & Matsuda, S. (2011). Legal cynicism, collective efficacy, and the ecology of arrest. *Criminology, 49,* 443-72.

Kleck, G. (1997). *Targeting Guns: Firearms and their Control.* New York, NY: Aldine De Gruyter.

Landsberger, H. A. (1958). *Hawthorne Revisited: Management and the Worker, Its Critics, and Developments in Human Relations in Industry.* Ithaca, NY: Cornell University Press.

Luca, D. L. (2015). Do traffic tickets reduce motor vehicle accidents? Evidence from a natural experiment. *Journal of Policy Analysis and Management, 34,* 85-106.

MacDonald, H. (2015). The new nationwide crime wave. Retrieved from:

http://www.wsj.com/articles/the-new-nationwide-crime-wave-1432938425

MacDonald, J., Fagan, J., & Geller, A. (2016). The effects of local police surges on crime and arrests in New York City. *PLOS ONE, 11*(6), e0157223.

McCain, L. J., & McCleary, R. (1979). The statistical analysis of the simple time-series quasi-experiment. In T. D. Cook & D. T. Campbell (Eds.), *Quasi-Experimentation: Design and Analysis Issues for Field Settings* (pp. 233-93). Chicago, IL: Rand McNally.

McCleary, R., & Hay, R. (1980). *Applied Time Series Analysis for the Social Sciences*. Beverly Hills, CA: Sage.

Mesa Police Department. (2013). *On-Officer Body Camera System: Program Evaluation and Recommendations*. Mesa, AZ: Mesa Police Department.

Miller, L., Toliver, J., & Police Executive Research Forum (2014). *Implementing a Body-Worn Camera Program: Recommendations and Lessons Learned*. Washington, DC: Office of Community Oriented Policing Services.

Morin, R., Parker, K., Stepler, R., & Mercer, A. (2017). *Behind the Badge: Amid Protests and Calls for Reform, how Police View their Jobs, Key Issues and recent Fatal Encounters between Blacks and Police*. Washington, DC: Pew Research Center.

Morrow, W. J., Katz, C. M., & Choate, D. E. (2016). Assessing the impact of police body-worn cameras on arresting, prosecuting, and convicting suspects of intimate partner Violence. *Police Quarterly, 19*, 303-25.

Newport, F. (2014). Gallup Review: Black and White Attitudes toward Police. Retrieved from: http://www.gallup.com/poll/175088/gallup-review-black-white-attitudes-toward-police.aspx

ODS Consulting. (2011). *Body-Worn Video Projects in Paisley and Aberdeen, Self-Evaluation*. Glasgow, UK: ODS Consulting.

Packman, D. (2011). *2010 NPMSRP Police Misconduct Statistical Report*. Washington, DC: Cato Institute.

Peterson, B. E., Yu, L., La Vigne, N., & Lawrence, D. S. (2018). *The Milwaukee Police Department's Body-Worn Camera Program: Evaluation Findings and Key Takeaways*. Washington, DC: Urban Institute, Justice Policy Center.

Peterson, R. D., Krivo, L. J., & Browning, C. R. (2006). Segregation and race/ethnic inequality in crime: New directions. In F. T. Cullen, J. P. Wright & K. R. Blevins (Eds.), *Taking Stock: The Status of Criminological Theory* (pp. 169-90). Piscataway, NJ: Transaction Publishers.

Police and Crime Standards Directorate (2007). *Guidance for the Police Use of Body-Worn Video Devices*. London, UK: Home Office.

Puente, M. (2014, September 28). Undue force. Retrieved from:

http://data.baltimoresun.com/news/police-settlements/

Pyrooz, D. C., Decker, S. H., Wolfe, S. E., & Shjarback, J. A. (2016). Was there a Ferguson effect on crime rates in large U.S. cities? *Journal of Criminal Justice, 46*, 1-8.

Schmidt, M. S., & Apuzzo, M. (2015, October 23). FBI chief links scrutiny of police with rise in violent crime. Retrieved from: https://www.nytimes.com/2015/10/24/us/politics/fbi-chief-links-scrutiny-of-police-with-rise-in-violent-crime.html

Schneider, B. (2015). *Do Americans Trust their Cops to be Fair and Just? New Poll Contains Surprises*. New York, NY: Reuters.

Sherman, L. W. (1993). Defiance, deterrence, and irrelevance: A theory of the criminal sanction. *Journal of Research in Crime and Delinquency, 30*, 445-73.

Suler, J. (2004). The online disinhibition effect. *Cyberpsychology and Behavior, 7*, 321-26.

Truman, J. L., & Langton, L. (2015). *Criminal Victimization, 2014*. Washington, DC: Bureau of Justice Statistics.

U.S. Department of Justice. (2015). *Justice Department awards over $23 million in funding for body worn camera pilot program to support law enforcement agencies in 32 states*. Retrieved from: https://www.justice.gov/opa/pr/justice-department-awards-over-23-million-funding-body-worn-camera-pilot-program-support-law

U.S. Department of Justice. (2002). *Police Use of Excessive Force: A Conciliation Handbook for the Police and the Community.* Washington, DC: Community Relations Service.

Weitzer, R., & Tuch, S. A. (2004). Race and perceptions of police misconduct. *Social Problems, 51*, 305-25.

White, M. D. (2014). *Police Officer Body-Worn Cameras: Assessing the Evidence.* Washington, DC: Office of Community Oriented Policing Services.

White, M. D., Gaub, J. E., & Todak, N. (2018). Exploring the potential for body-worn cameras to reduce violence in police-citizen encounters. *Policing: A Journal of Policy and Practice, 12*, 66-76.

ABOUT THE AUTHORS

Lisa Stolzenberg is a professor and chair of the Department of Criminology and Criminal Justice at Florida International University's Steven J. Green School of International and Public Affairs. She is also president of We Love Animals Rescue, Inc., a 501(c)(3) public charity located in Hollywood, Florida. She has a BA in criminal justice from the University of Florida and an MS and PhD in criminology from Florida State University.

Stewart J. D'Alessio is a professor in the Department of Criminology and Criminal Justice at Florida International University's Steven J. Green School of International and Public Affairs. He previously served as a captain in the Military Police Corps and participated in Operation Just Cause and Operation Desert Storm. He has a BA in history from Stetson University and an MS and PhD in criminology from Florida State University.

Jamie L. Flexon is an associate professor in the Department of Criminology and Criminal Justice at Florida International

University's Steven J. Green School of International and Public Affairs. Her research interests include juvenile delinquency, juvenile psychopathy, criminal justice issues related to punishment, and policy evaluation. She has a BA, MA and PhD in criminal justice from the University at Albany, State University of New York.

www.ingramcontent.com/pod-product-compliance
Lightning Source LLC
Chambersburg PA
CBHW041221270326
41933CB00001B/2